*A special dedication to Graham Victor Parker, the Frogfather*

I was on my way to school today when I stopped to cross the road

What I saw took me by surprise,
"Holy Mary Mother of Toad,"

A big white van drove past me as I did a double take,
My big frog eyes were still in shock, surely this was fake.

The writing on the van was the colour of my skin,
And on the back and sides it said frog in humongous lettering.

Underneath frog
I think the smaller words said,
Diamond drillers, I'm not too sure but I think
that's what I read.

FROG
DIAMOND DRILLERS

I went to school amazed, when a teacher
came up to me,
Good morning Freddie, how are you?
So I told her what I see.

Oh Freddie you are funny, your imagination's wild,
Frogs don't drill for diamonds she said as she looked at me and smiled.

Later on that day I was sitting in my class, The teacher got up from her desk and looked at us and asked,

"Today I want to talk to you, I have a question for you all,
What do you want to be when you're older, once you've left this school?"

I put my hand up high, I couldn't wait to say, "I want to drill for diamonds," I said, I see a van today.

"oh Freddie, you're a dreamer, that's just not what frogs do."

"oh teacher you are wrong," I said, I know that that's not true.

"There's a company called Frog, they are diamond drillers you know?"
"I'll get a job with them and drill for diamonds, rain, sleet or snow."

I rushed straight home from school and asked to use the phone,
I found Frog's number on the net and went upstairs alone.

"Hello thanks for calling frog, my name is Shell, how are you today?"
She sounded really nice and friendly, I told her I'm okay.

"I'd like a job with frog" I said,
"but I'm only ten."
"Come down here on Saturday,
you can go to work with Ben."

I couldn't believe my ears as I thanked Shell for her time,

And dreamt that on Saturday I'd be down a diamond mine.

On Saturday I woke up early
I really couldn't wait,
I left my house and hopped on down to
Springhead Industrial estate.

We didn't drill for diamonds and there was no diamond mine,
Instead we used diamonds to drill, it blew my froggy mind.

We cut through loads of concrete, with a diamond tipped blade
What a waste of diamonds I thought, imagine all the engagement rings that we could've made.

They offered me a job, to which I told them, "Cheers."

I start for Frog when I'm sixteen, in 6 Froggy years.

www.ingramcontent.com/pod-product-compliance
Lightning Source LLC
Chambersburg PA
CBHW051836210526
45473CB00005B/1899

# DIAMOND ELITE MAGAZINE

**2ND QTR 2018**

*"A Different Type of Magazine"*

## Content:

Pg.2 Patrina Dixon
Pg.3 G. Dundas, Sapphire Chic, GeeChee Vintage and Jafra
Pg.4 Sheila the Credit Boss & Atiya McNeal
Pg.5 Black Lyfe Publications
Pg.6 Norma's | Kami Redd | Diva Cosmeticz | Scentsy
Pg.7 Mother's Retreat | My Beautiful Fluff | Sweet Surrender
Pg.8 Sensationally Nappy | Say It Witcha Chest
Pg.9 Mixxed Marketing | Glamorous Trois Hair
Pg.10 3JsCreation Production Company
Pg.11 Posh Kingdom Clothing | Credit Repair | Donna Earl
Pg. 12 Meal Prep
Pg. 13 Cherish | It Works!
Pg. 14 Mark Edwards' Interview
Pg. 15 The Best Poet You Know
Pg. 16 Primerica | 4G Realty Group | Greatness Pursued
Pg. 17 TLC | Platinum Pleasures | Travel | Sister Sankofa
Pg. 18 Running Your Race LLC
Pg. 19 Consult Positivity LLC
Pg. 20 Project Resiliency | Seed to Seeds
Pg. 21 Kamelda's | Paparazzi | Beauty of Credit | Therapeutics
Pg. 22 Featured Carriers
Pg. 23 Testimonials

## Mission Statement:

Diamond Elite Magazine's goal is to boost the exposure and sales of entrepreneurs. We believe networking and word of mouth are the biggest essentials when it comes to small business. As the years continue, we plan to thrive in success and help expand the small businesses who have contributed along the way.

**Author Mark Edwards**
Diamond Elite Magazine is truly about supporting, uplifting and ensuring that their clients are properly advertised... Nothing but love for this magazine company!!

Be Sure to Take a Photo and Contact the Small Business Owners/ Entrepreneurs directly if you would like to purchase a product or service!

# PATRINA DIXON

### Award-Winning Author | Speaker | Certified Financial Education Instructor

(860) 607-3275
patrina@itsmymoneyjournal.info
www.itsmymoneyjournal.info

 @itsmymoneyjournal

 @itsmymoney_

Patrina Dixon, Certified Financial Education Instructor and award-winning author of the top-selling financial guided series, "It'$ My Money™" is an advocate for financial literacy. She is also a member of the Black Speakers Network and a 100 Women of Color Honoree for 2018. Patrina has a passion for serving her community and uses her company, P. Dixon Consulting, LLC to offer money management strategies to people of all ages. She led Hartford's Financially Fit community event and enjoyed her role as a Personal Finance volunteer for Junior Achievement. Patrina is shaping the spending and saving behaviors of her clients with a goal of guiding them toward financial independence.

Through her education received at the University of Hartford's Barney School of Business, as a proud mom and a life-long financial advisor – she understands and thrives at teaching the importance of financial literacy. Patrina also holds a Financial Management Certificate from Cornell University. The It'$ My Money™ journal series, workshops and book clubs allows Patrina to educate and enlighten families on their finances. She is dedicated to molding the next set of financial leaders. Patrina is a devoted wife and mother who resides in Connecticut.

**Available for speaking engagements, facilitating financial workshops, one on one financial coaching, and credit restoration.**

## Published Works

**It'$ My Money ™ Journal** is the first in a series of journals dedicated to financial literacy and money management. This book is designed to enrich teenagers with simple but smart financial information while providing inspiring quotes and guided questions.

**Entrepreneurship: My Story, Your Guide** is a collection of stories and experiences from entrepreneurs in various industries across the United States. Learn from these successful entrepreneurs some of the exact steps they took to achieve success in their industry.

## Media & Partnerships